NELSON'S
HORSEMASTER SERIES

Rearing a Foal

JANE HOWELL

Rearing a Foal

Thomas Nelson & Sons Ltd.

THOMAS NELSON AND SONS LIMITED

36 Park Street London W 1
PO Box 27 Lusaka
PO Box 18123 Nairobi
PO Box 21149 Dar es Salaam
77 Coffee Street San Fernando Trinidad

THOMAS NELSON (NIGERIA) LTD
PO Box 336 Apapa Lagos

THOMAS NELSON (AUSTRALIA) LTD
597 Little Collins Street Melbourne 3000

**THOMAS NELSON AND SONS (SOUTH AFRICA)
(PROPRIETARY) LTD**
51 Commissioner Street Johannesburg

THOMAS NELSON AND SONS (CANADA) LTD
81 Curlew Drive Don Mills Ontario

This book first published 1970
© Jane Howell 1970
Printed in Great Britain by
Butler & Tanner Ltd Frome and London
SBN 17 147205 5

CONTENTS

ILLUSTRATIONS

(between pages 40 and 41)

Chapter 1

PREPARATION

Preparations for the rearing of a foal begin when the decision is made to send a mare to stud. To be successful in this venture needs an understanding of the problems involved and a realization that both time and money will have to be expended. Owners must ask themselves 'Is the available accommodation suitable?' and 'Is the grazing area large enough to accommodate two animals?' The foal will grow up and all to soon require the same amount of space as the mare. If the answer to either question is no, ensure that the matter can be remedied before embarking on an addition to your equine family. It is amazing how problems can be overcome if they are recognized as problems in good time. Remember, too, that once the decision is taken to breed a foal at least four years must elapse before the horse or pony is sufficiently developed for riding. Ideally, if these homebred animals are to have a long, active life, they should not do any serious ridden work until they are four, and to ride them at two years old amounts to cruelty, since the bones have not yet fully set. At this age horses also lack the required muscular

development and can suffer considerable pain, and sometimes irreparable damage to the spine and joints, if they
are ridden. It is, indeed, rather like asking a ten-year-old
child to carry hundredweight sacks of corn on its back.
Patience must, therefore, be an attribute of would-be
breeders, although, riding excepted, there is still much to
be done during the early days to lay the foundation for a
sound and sensible mount in future years.

Selection of a suitable stallion is usually made some
months before the beginning of the breeding season. A
mare owner will probably have visited several studs before
arriving at the final choice, following which the stud owner,
or his manager, will have been asked to accept the mare
for service. This is known as reserving a nomination. Every
stud has its special terms of business and mare owners
should enquire about these. Expenses to be expected, over
and above those for the keep of the mare and the stud fee,
are veterinary fees, charges for farriers' attention, cost of
transport for the mare to and from stud and usually a
groom's fee, about £1 or £2, handed by the stallion owner
direct to the man, or nowadays to the girl, in charge of the
horse. It is left to the mare owner to send or give something in appreciation of the work done by the other stud
hands. The everyday care of your mare will be in the
hands of these people, who are hard working, conscien-
tious and genuinely anxious that she should settle down
quickly on arrival, and in due course return to you fit, well
and in foal. There is also considerable paperwork for the
person in charge of the office, and prompt payment of stud

bills is appreciated. These bills nearly always include farriers' charges, but bills for veterinary treatment usually
come direct from the veterinary surgeon concerned.
Charges for keep vary and depend on whether a mare is
either at grass without extra feeding, at grass with extra
feeding, or stabled at night. Only very rarely are mares
kept in all the time, since brood mares need exercise and
good grass to encourage the breeding cycle, although it
may be necessary to keep a mare with young foal at foot
stabled if the weather is very bad. If a mare at stud has a
foal at foot the charges for keep are usually in the region
of 50 per cent extra in all categories.

It is helpful to the stud concerned, and paves the way to
good future relationships, if you ascertain from them that
the day upon which you intend to send your mare to stud
is suitable, and if at the same time you advise them of the
anticipated time of her arrival. Tell them, also, of any particular points about your mare: for instance, whether she
is sometimes difficult to catch, whether she is easily
'bossed' in the field by other horses, or whether she herself is inclined to be 'bossy'. You should also let them
know whether the mare has a tetanus toxoid immunization;
accidents can happen on the best-run establishments and
tetanus is a killer disease from which fortunately it is possible to protect horses. Tetanus antitoxin *can* be given at
the time of an accident but it only protects the animal for
21 days. Before a mare departs for stud, attention should
be given to worming and to the rasping of teeth and veterinary advice should be taken regarding these two items.

Most mares are sent to stud without their shoes, and in any event hind shoes *must* be removed. The stud will arrange for the mare's feet, and shoes if any, to have regular attention at intervals of about four to six weeks while she is with them. A comfortably fitting headcollar should be worn by the mare and a refinement is to have a brass plate attached to it bearing her name.

The gestation period in the horse is 11 months, although some may carry their foals for up to 12 months and others for as little as $10\frac{1}{2}$ months. However, when calculating the expected date of birth of a foal one must take the average, and a mare covered in mid-May can be expected to foal about mid-April the following year. Mares come into season at intervals of 21 days and remain in season for periods varying from three to seven days, during which time they will probably be covered every other day. If a mare is to be manually examined for pregnancy before returning to her owner – and this is a very wise practice – she will need to remain at the stud for 42 days after covering so that this may be done. Of course, if a mare is covered and then comes into season three weeks later she will have to be covered again, and the 42-day period counted from the second covering date. If this practice is followed a mare will be away for a minimum of six weeks, even supposing that she is in season on arrival at the stud and holds to the first service. If she has been in season shortly before going to stud the length of time away could be eight or nine weeks and it will be more if she does not hold to the first covering period.

In the British Isles the months from mid-April to mid-June may be expected to offer the best weather and good grass, although in the Northern areas the seasons are somewhat later. Therefore, owners should endeavour to arrange for mares to foal during this period if possible. Foals are, of course, born outside this two-month period and not only survive but do well, although owners must be prepared to offer suitable accommodation for the early foals, since the weather can be very unpleasant in February and March, whilst the later foals will need extra attention to their feeding as the best grass has usually gone by the end of June.

While the mare is at stud the preparation can go ahead for her winter care and the well-being of the expected foal the following spring. A start can be made by purchasing a foal slip from your saddler, who will know the size to give you if you tell him the size of the mare and also the size of the stallion she has visited. If you do not already have some buy also some Neatsfoot oil. Thoroughly oiling the foal slip with this preparation will protect the leather from the weather and will also make the new leather soft, pliable and more comfortable for the delicate skin of the foal, who will be wearing this slip from an early age. This advice may sound a little like counting one's foal before it is born, but it is something which must be done if the new-born foal is not to receive a badly rubbed face from brand new and unsoftened leather. As an alternative to leather, foal slips made of webbing can be bought. These are certainly very soft, but if a webbing slip is adjusted to fit comfortably when it is dry it is likely to shrink when the foal is out in the rain and will then become too tight. To fit any foal slip too loosely is not wise because foals have a habit of scratching their heads with a hind leg and it is possible for a leg to get caught up in the slip, when the result might be a broken neck or limb.

Next turn your attention to the stable accommodation. It should be spring-cleaned and disinfected and checked in case any possible sources of danger have escaped your notice. Since horses are basically not thinkers, and where their own safety is concerned are incredibly stupid, flight being their first line of defence, a foal needs every possible protective care from the moment of its birth. Even if the mare foals outside, and it is desirable that she should do so unless a very large and suitable box is available, it may be necessary to bring mare and foal into the stable quite soon after birth, so make sure that their are no sharp protrusions, such as nails or wood splinters, on which the foal might catch itself or its foal slip. Mangers of a square or oblong type placed in the corner or along one wall present hazards because of their very sharp corners. Corner type mangers are best perhaps best, and those with an anti-spill rim are excellent for saving food. Windows should be reasonably high and the glass protected by bars or suitable wire mesh. The day will come when it is necessary to leave the foal alone in the box, and it will, at first, undoubtedly try to get out. Unless these precautions are taken a nasty accident could be caused by the foal putting a foot through a window pane.

Doors should open outwards and there must be no gap between the lower edge and the floor. The danger here is that when the foal is lying down it may slide one of its very small feet under the door and could easily break a leg when it attempts to rise.

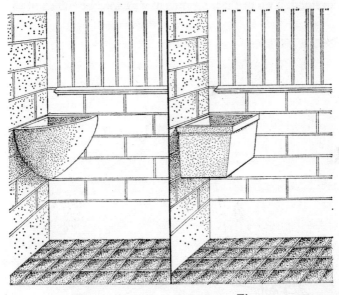

Fig. 1.   *Fig. 2.*

Fig. 1. A corner manger which presents no dangers in the way of a sharp edge.

Fig. 2. A square-type manger. It can be seen how easy it would be for a foal to be injured on the sharp edges.

Attention must also be paid to the paddock and its fences. Ensure that the latter are in good order and if made of wire that each strand is really tight. Most people have to use wire fencing these days because of the high cost of

7

posts and rails, and on the whole it is quite safe, providing it is taut. It is the slack strand of wire, through which an animal can put a head or leg, which causes the damage. Therefore, see that all fencing posts are firmly in the ground and that the wire attached to them is properly strained. Two or three strands of wire may be sufficient to

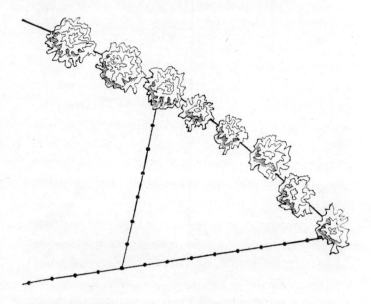

Fig. 3. An aerial view of the way fencing should be used to eliminate the dangers of a tight corner in a grazing field.

contain a quiet mare but four are a much more sensible proposition where foals are concerned. Often in the three-strand fence the lower strand is of a height perfectly suitable to contain the foal whilst it is on its feet but is not low enough to prevent the youngster who has lain down very close to the fence from finding itself in the next door field when it gets up. Very often foals will roll before rising and that is the time when they are most likely to go under the lower strand. This can be particularly dangerous if the fence is one bordering a stream or pond, for once the foal is through the fence it can easily enough fall into the water. Such an accident could well bring about the death of a new-born foal, who at the time will not have sufficient strength to struggle out.

Advantage can be taken of the mare's absence at stud to carry out any maintenance required with regard to grazing land. Few owners of horses and ponies have sufficient grazing and as a result their pastures are frequently sour. The mare's absence affords an ideal opportunity to rest the land, if possible to run cattle or sheep through it and to attend to weed control. Indiscriminate spraying is not advisable, since the horse requires a proportion of such weeds as dandelions and narrow rib plantains, but there are a number of undesirable and very often persistent weeds which one should make every effort to eradicate. Advice concerning weed control is easily obtainable from the local agricultural advisory officer. Stinging nettles, a common and very persistent weed, can be controlled by frequent cutting.

A constant supply of clean fresh water is essential for the well-being of the mare and her foal, and if this is not laid on to a suitable water trough in the field a container must be provided. Buckets are not ideal as they are easily tipped over and present a hazard if mare or foal should get a foot through the handle. A good and safe water container

is an old porcelain bath, which can sometimes be obtained
very reasonably from a local builders' yard. These porcelain baths hold a very adequate quantity of water, which
can be supplied to them by hose, and they are easily cleaned.
They must be positioned either flush against a fence or
well away from it, but never in such a position that a narrow space, into which a foal would most certainly manage
to squeeze itself, is left between the bath and the fence.
It is also essential to ensure that there are no right-angled
corners in the field. If there are they must be boarded
across so that a galloping foal is persuaded to turn without
running into the fencing (see diagram). If the field posses-
ses a tall hedge against cold winds this is ideal, but, of
course, no hedge must contain anything of a poisonous
nature, such as yew, which nearly always kills horses and
ponies even if it is eaten in very small quantities. If a field
shelter is provided this must receive the same meticulous
care with regard to danger spots as did the loose box. It
will also be necessary to treat all the woodwork with a suitable preservative to protect the timber and discourage the
foal from gnawing, a habit which is easily acquired and
very difficult to cure.

While the mare has been at stud you will have been advised
by the stud owners from time to time of her progress: such
as when she was covered, if she held or if she returned to
the stallion again three weeks later, and so on. You may even
have been to see her, although frequent visits of owners
to see their mares are not viewed with any great enthu-
siasm by stud proprietors who, together with their staff, have

a great deal to do during the season. In any event it is a cardinal sin to visit without giving prior notice of your intention to do so.

Should you be told at the end of the season that your mare is not in foal do not be too disappointed. A percentage of mares return from stud every year in this condition and it is probably not due to any particular fault either on the part of the mare herself or of the stud concerned. Throughout the world the percentage of mares who do not conceive in any one year is unfortunately still quite high, and it is possible that you will be one of the owners who has to face this disappointment.

Chapter 2

CHOOSING YOUR STALLION

There are three important considerations to be taken into account when selecting a stallion. The first is that bad points in the potential brood mare – a poor hind leg, for example – should as far as possible be eliminated by the choice of a stallion who is particularly strong where the mare is weak. Secondly, you should think carefully not only about the stallion himself, but also about what sort of animal you want to produce. The third point, which is equally important, is the question of temperament. Obviously, if your mare tends to be rather excitable and highly-strung it is wise to choose a stallion who is placid and kind. Gaiety and high spirits are quite permissible in a horse but bad temper is not. In fact, many people consider that bad temper in the mare is more dangerous. There is not only the risk of injury to the foal, but because it will copy its dam it may, at an early age, pick up her bad habits. Fortunately, in-bred bad temper in horses is rare; most ill-tempered animals have good reasons for behaving unpleasantly.

If the mare is a registered native breed and a pure-bred

foal is required it is necessary to send her to a registered stallion of the same breed. Application to the breed society will usually result in information regarding registered stallions at stud during the current year. This information will probably take the form of a stallion list, which will contain the names of the stallions and the names and addresses of owners to whom application for further details should be made. A list of breed and other societies, with their addresses, will be found at the end of this book. In addition to supplying details of available sires each society will advise owners of conditions for the registration of foals, from which it will be seen that should an owner possess a mare about whom no breeding details are available it is still possible, by the selection of a suitable sire, to breed a foal which will be eligible for one or other of the various part-bred registers.

Apart from the native breeds there are thoroughbred Arab and Anglo-Arab sires from which to choose, and owners of larger mares may use one of the sires under the Hunters' Improvement Society's Premium Stallion Scheme. This society, generally referred to as the H.I.S., with the help of the Horse Race Betting Levy Board, award in the region of 65 premiums and upwards of a dozen super premiums each year at a cost of over £30,000. Under this scheme members of the H.I.S. may send non-thoroughbred mares to visit one of the very excellent stallions for a fee of 12 guineas. Non-members must pay 22 guineas, and in each case there is a groom's fee of 10 shillings. These fees in themselves would, of course, be hopelessly uneconomic for the stallion owner under normal private circumstances. They are made possible by the support provided through the society. Every owner of a stallion which is awarded a premium receives a grant in the region of £450, which is in effect a subsidy and enables him to use his horse at the otherwise impossibly low stud fee of 12 guineas. In return the owners undertake to make their stallions available in prescribed districts. Winners of super premiums receive additional awards ranging from £80 to £270, but within the workings of the H.I.S. Scheme, premium and super premium winners stand exactly level, meaning that all these sires are available at the above-mentioned fees to the owners of non-thoroughbred mares, allowing them the benefits of sires who would otherwise stand at fees of perhaps £48, or even £98. Those who own thoroughbred mares and wish to use H.I.S. sires are free to do so, but the fees must be arranged privately. Thoroughbred matings fall outside the scope of the society's arrangements. H.I.S. stallions are allocated on a county basis and travel their area during the season. This means that most owners will be able to find a suitable H.I.S. sire within easy distance of their homes. The season begins on 13th April and continues until the end of July. The society also sponsors a Brood Mare Premium Scheme at affiliated shows for which mares producing foals by stallions which have been awarded Premiums in the year of service are eligible to compete.

Except in certain circumstances all stallions of two years

old and over have by law to be licensed by the Ministry of Agriculture. They are divided into two groups, pedigree and non-pedigree. The latter is to some extent a misnomer, since many stallions in this group have known breeding of several generations on both sides but are not pure-bred, often being part-Arab, part-Welsh, part-Dartmoor and so on. Owners are assured that stallions holding a licence are sound and free from hereditary diseases.

Certainly in the pony section, size will be of importance. There is, unfortunately, no hard and fast rule which will guarantee that a foal from parents of particular heights will mature to a standard size, and the fact that a mare of 12.2 h.h. visits a stallion of 14.2 h.h. does not mean that the resultant foal, when fully grown, will be 13.2 h.h., although this can, and does, sometimes occur.

All inherited characteristics are governed not only by the immediate parents but, to a lessening degree, by the grandparents and great grandparents; how often one sees a tall child whose parents are of medium height and is told by the parent 'of course my father (or mother) was tall'. Therefore, when considering breeding stock with regard to size one must, as far as possible, study the size and type of their antecedents. Type is important, since nothing is a greater misfit than a pony with, for example, the head of a hunter. If a pony mare is of unknown breeding there is nothing to be done on that side and one is faced with choosing what appears to be the most suitable stallion. In doing so it is as well to make allowances for the improved feeding and management that present-day conditions offer and to err on the side of 'under' rather than 'over' height; better a 14.1 h.h. pony in a 14.2 h.h. class than a 14.3 h.h. animal who is only eligible for hunter classes but is of the pony type. Most difficult to assess are those animals who, although entered in a part-bred register, are included there as a result of a sufficient concentration of one particular breed from one side only; the unknown side may carry factors such as thoroughbred or hunter blood. These factors, or any one other factor, could appear in the progeny. For instance, there could be a tendency for the foal to outgrow one's expectations by a considerable amount. It is for this reason that 'type' as well as size should be considered when viewing potential breeding stock.

Within the various native breed registers there are permissible height standards laid down and so, assuming that an owner has a registered mare of a certain breed and wishes to produce from her a foal who will mature above this particular breed's height, he or she may look to, perhaps, an Arab or small thoroughbred sire. It has been shown that in nearly all cases where a mare has been mated with a stallion considerably larger than herself, both in height and build, she will produce a foal who at birth is not disproportionate to her own size, but who will undoubtedly be larger than its dam when fully grown. There is, therefore, not too much to worry about with regard to the mare's well-being at foaling time on this score. In the not very likely event of foaling complications arising out of such cases it is probable that they would have occurred even had the mare visited a stallion of her own size. Assuming that a mare of 13.2 h.h. visits a suitable and carefully chosen stallion of 15 h.h., the resultant foal might well mature to 14.1 h.h., with a very fair chance that it would stay within the usual class limit of 14.2 h.h. At this height a certain degree of thoroughbred quality is permissible, although it is not so acceptable in the small classifications where true pony type is required among the entrants. A 12.2 h.h. animal may be crossed with one of 14 h.h. and one may reasonably hope for the foal to grow to 13.1 h.h., or even an inch larger. A great deal depends on the background of the 14 h.h. animal; is it pony, cob, or thoroughbred in type? If the latter, what size were its parents? All these are points to consider, for if the animal was of a larger type there is the chance that one may end up with a quality pony type of 14 h.h.

Quite a number of people have very strong feelings with regard to colour, but by and large the tales told about a particular colour being prone to certain good or bad characteristics are mostly based on isolated experience. For instance, the number of really black horses is small and therefore should one of these at some time prove to be difficult to manage it will be remembered, since the colour is not common. Likewise, hoof colour has been a subject for discussion, many persons insisting that white hoof is weak hoof, yet there are good feet that are white: one has only to think of the many Arabs with white socks or stockings, who as a breed are not noted for poor feet, to realize that the belief is not well-founded. Undoubtedly poor white hoof does exist, but so does black hoof of poor quality, and one is forced to conclude that inferior hoof is, like other failings, probably an individual physiological condition.

Having selected the stallions most near to one's requirements it will be necessary to arrange a visit to the studs where they are resident. You should make an appointment with the owner or stud manager before you go, and endeavour to arrive on time. Such pre-arranged appointments are not to enable the stud to hide anything, but to allow time to be set aside for the visitors, who can then be made welcome. Visits without prior appointment are not popular and although, except on rare occasions, visitors will not be turned away, they may well find that it is not possible to see all they, or the stud owner, would wish. Both in and out of season stud farms are extremely busy places, and careful planning is needed if all the many necessary duties are to be carried out each day. An unexpected visitor will not have been included in the plans and will thoroughly disorganize the routine with even a short visit. Stock which could have been held in the boxes pending a visitor may well have been turned out and will have to be fetched in again. This in turn will mean that beds will have to be laid, since they have probably been 'put up' to allow the floors to air.

On arrival make yourself known and await the owner or his representative. One of the cardinal sins of the equine world is for a stranger to wander into a stable uninvited.

When fairly considered, this is not surprising. One does not, after all, walk unannounced into other people's houses.

At the outset the horse will be seen in his box, and here his temperament will be noted. First impressions are nearly always right and if he is quiet, and kindly disposed to his handler, he is probably pleasant in other ways. Do not enter the loose-box without either being invited or asking if it is permissible. Following this introduction the horse will be led out and stood, so that a good view may be had from all angles. When moving round him do so quietly and at a safe distance; it is amazing how much space he will need if something surprises him or if he is feeling a bit gay. When he has been seen at the halt he will be trotted out so that his action may be noted, and it is quite in order to ask to see him run up a second time. If the weather is bad it will be appreciated by all concerned if this outdoor session is not prolonged for a greater time than is necessary, and in any event one should remember that the handler will have other duties awaiting him; a smile and a 'Thank you so much' will indicate that he may return his charge to his box.

Ask to see some young stock by the horse, and if possible their dams as well, as this will give an idea of what may be expected from different types of mare. Bear in mind the faults of your potential brood mare and endeavour not to send her to a stallion with similar shortcomings. Try to avoid too much divergence in type. If a mare lacks substance by all means select a sire with plenty of bone, but see that both have a reasonable type resemblance. An attempt to increase size may, of course, be made by sending a small mare to a larger stallion. In normal cases the permissible margin is great and a mare of 12.2 h.h. may be covered by and safely produce the offspring of a stallion two hands larger, but again similarity of type must be considered. Careful perusal of the young stock and their dams will have indicated what may be expected from a particular sire when mated with mares a great deal smaller than himself.

While visiting a stud take note of the general appearance of the establishment; is everything neat and tidy and are the staff of a similar appearance; are the stock in reasonable condition, contented and friendly? If there is seen among the inmates just one horse or pony in poor condition there is no cause for alarm; the animal may have been ill or is perhaps that problem creature a 'bad doer'. Even among the human race there are some thin specimens who, for one reason or another, no matter how well fed they may be, are never anything else but bean-poles!

Having completed a tour of the establishment the business side of the affair must be discussed. Studs vary with regard to conditions upon which mares are accepted, and usually the more precise these are the greater will be the efficiency with which the place is run. Efficiency is well worth paying a little more for, since in the long run it may save several weeks of keep charges. The amount of the stud fee will be quite clear, but payment methods vary. The whole fee may be payable at the time of reservation of the nomination, in which case there will most probably be a 'no foal, free return' clause. This simply means that any mare who does not produce a foal will be covered without further payment the following year, subject to the stallion being available and to the mare owner having produced a veterinary certificate of barrenness by a stipulated date, usually 31st October in the year of covering. The mare will not, however, be kept free of charge in the second year. All such charges and also veterinary fees, if any, will be the responsibility of the owner as in the first year. Alternatively, the fee may not be payable until the mare is settled in foal. This is a 'no foal, no fee' clause, and if the mare is not examined for pregnancy while at stud the owner must produce a certificate of barrenness by a stipulated date or be liable for the stud fee.

Keep charges may at first seem high but it must be realized that endless work is entailed and while an owner who does his own animal at home will not think in terms of his own labour costs, a well-run stud must do so, as it has to pay wages. It may be as well here to consider just what work does go on and the labour and responsibility involved. It is probable that every mare at grass has a minimum of half an hour spent on her each day. Not only does she have to be seen in the paddock twice daily as a routine measure to ensure that all is well with her, but must be brought in to be tried or covered as well. Until she is covered she may well be brought in every day and this may have to be carried out for two weeks or more if

she has been in season just prior to arrival. It is not difficult to appreciate how many man hours per week have
to be expended upon her, before her actual cost of keep
has been considered.

Even if the mare is not having extra food there will be,
on a well-run stud, a worm control policy, which entails
giving a feed once a week in which is included the necessary worm dose. Here again, the mare must be fetched
from the field, put in a box, which for safety reasons must
be bedded down, and, when she has been returned to the
field, the box must be tidied again. Admittedly the box,
when bedded down, will probably be used for a number
of mares in succession, but the time factor involved in
fetching in and putting out is the same for each animal.
This is of course part of the business of a stud farm, but
all such things have to be accounted for in arriving at
costs, and owners will realize why keep charges cannot be
low. The object of sending a mare to stud is that she will
have a foal in the following spring and owners must be
prepared to spend a reasonable sum of money in order
that this should happen.

Chapter 3

PRE-NATAL CARE

In-foal mares returning from stud, after approximately two months of idleness, will be soft and in some cases over-fat. Where pony mares are concerned the return frequently coincides with that of the children from school, and the latter will no doubt be anxious to ride. There is no reason why an in-foal mare should not be ridden but, since the object of the exercise so far has been to get a mare in foal, discretion must be exercised with regard to work and the companions with whom she is turned out. If it is at all possible, do not keep the expectant mother in a paddock with a gelding, as his attentions can cause a mare to slip her foal, and on no account should she be permitted to run with a colt. Better by far to arrange for the mare to have female companionship or to live by herself, if she will do so without trying to get out. If you do not have another mare or filly, and your mare will not live alone contentedly, try to find another owner with an in-foal mare and let both mares live together. Another advantage of doing so is that both owners will be glad of the companionship that the foals will give each other later on, after they have been weaned. If the mares do not know each other delay having hind shoes put on until they have lived together for a week or more, so that if there is a kicking match there will be less likelihood of serious injury. At no time should the hind shoes have studs inset since these can cause serious wounds.

When starting to ride the mare after her period away care must be taken to keep a watch for any signs of girth galls or a sore back. The former will be more likely to appear as pregnancy advances, when the saddle will be pushed forward by the steady increase in the size of the mare. This is particularly true of short-backed mares. Daily applications, during the first week, of surgical or methylated spirits on the area behind the elbows will help to harden the skin. The same treatment may be given to the bars of the back under the saddle; that is, to the layer of muscle on either side of the spine.

Exercise during pregnancy is essential, but it should not take the form of anything more violent than gentle hacking. Show jumping and other strenuous activities are best avoided as the effort entailed could cause the mare to slip her foal. The same consideration applies in the case of gymkhana events. Owners and their children must decide just how much risk they are prepared to take, but it is well to remember that if you want a foal you cannot have your riding exactly as in other years.

This may be hard to accept when one thinks of conditions in the wild, where mares gallop over rough country with the herds every day, even when well advanced in pregnancy, and still produce foals. A little more thought, however, will cause one to realize that these wild horses and ponies lead a much tougher life than their domestic counterparts, covering, on most days, more miles than the average family mount covers in a week. Wild horses and ponies don't have to carry a human burden or as much excess fat as most domesticated equines, who, in most cases, lead a life of alternating extremes, being over-fed and under-exercised, as, for instance, in the term time when young owners are away, and then being put to sudden and vigorous work on their owners' return, often without sufficient preparation and sometimes, I regret to say, without proper food.

Just as in the previous chapter a careful look had to be taken at the space and facilities available, so now owners must decide whether or not they are prepared to give up some of their riding in the interests of breeding a foal. Riding is possible for at least the first four months of pregnancy, but after this time some mares will become uncomfortable and also sluggish. Most of them give a reasonably comfortable ride for the first five or even six months of pregnancy, but then it is time for the shoes to be removed and the mare allowed to get on with the business of producing her foal. In the case of children's ponies this really means that they will get their summer riding nearly as normal, perhaps a little riding at Christmas but, of course, none during the Easter holidays.

By the end of August the best of the grass, as far as its feed value is concerned, will have gone, and by September additional feeding must begin. One of the easiest methods of supplying such feed is in the form of horse and pony nuts. There are many brands from which to choose and makers are always pleased to provide details of their products, some of which are especially designed with breeding stock in mind.

At the beginning of the winter feeding quantities need not be large but they must be given regularly. In the case of small ponies, who have a tendency to be exceptionally good doers, extra feeding may be delayed until the beginning of October provided there is no sign of loss of condition. Each horse or pony must, of course, be treated as an individual with regard to appetite and the following table is only a suggested guide for the various sizes.

Up to 12.2 h.h. commencing with 2 lb. increasing to   5 lb.
12.2–13.2 h.h.   commencing with 3 lb. increasing to   6 lb.
13.2–14.2 h.h.   commencing with 3 lb. increasing to   8 lb.
14.2–15.2 h.h.   commencing with 4 lb. increasing to 10 lb.
15.2 h.h. and
   over   commencing with 5 lb. increasing to 12 lb.

As the daily ration is increased variety must be added to the diet and good bruised oats, broad bran, etc. should comprise part of the total daily ration. Bran may be fed as approximately 20 per cent of this daily ration and should always be damped before being mixed with the other feed ingredients as it is a light, flaky food and can cause a horse or pony to choke. The addition of sliced apples, vegetables, etc. will give added variety, and the introduction of boiled linseed once or twice a week will be very advantageous. To prepare linseed for feeding, allow 2 oz. of seed to 1 pint of water. Bring it to the boil, stirring constantly, and then let it simmer in a slow oven for a minimum of three to four hours. The mixture on completion of its cooking will be very liquid, but will turn to jelly as it cools. Depending on the size of the mare, half to one pint of the jelly and seeds may be added to a bran mash. Good hay must always be available and in as much quantity as each individual will consume, while a salt lick, or a lump of rock salt, should always be available in either paddock or stable. Unless recommended by a veterinary surgeon additives are unnecessary, although where there is a distinct mineral shortage on the soil, as occurs in some areas, there will be a need for them. A word with the agricultural advisory officer is the best way to find out what the mineral condition is likely to be in any particular territory.

The period between December and March imposes the greatest strain on an animal's system. The feed value of grass is almost nil at this time of year and in-foal mares will be drawing heavily on their bodily reserves to maintain their own body condition and to provide the necessary requirements for the laying down of the unborn foal's structural development. Throughout January and February, in particular, a careful watch must be kept and note taken of any change in condition of the expectant mothers, the majority of whom, if provided with sensible rations in sufficient quantity, will do very well and give no cause for concern. Those animals which are known to be rather poor doers may need to have additional rations, in which may be included small amounts of flaked maize or other heating foods.

Clearly, it is sensible to put food to its best use, and to do so it is necessary to make sure that mares are not carrying a heavy worm burden. In other words, there is no point in wasting good food on worms.

The usual way to ascertain the degree of worm infestation is to send a dropping sample to your veterinary surgeon who will arrange for a worm count to be taken and will advise you what, if any, treatment is necessary. I cannot stress too strongly the dangers of worm infestation where grazing facilities are limited, particularly where it is not possible to give a rest period to paddocks or to inter-graze with cattle or sheep. All horses and ponies have worms and it is merely a matter of ensuring that these are controlled. A regular worm control policy is the best method of dealing with the problem and it may be based either on a small weekly dose or a larger dose given at two-monthly intervals. Owners must consult their veterinary surgeons and decide upon a policy that best suits their individual programme.

Attention must also be paid to the care of feet, since as pregnancy advances greater strain is put upon them and the lower limbs. To keep them in order it should be sufficient if the farrier makes a regular call once a month. Mares with poor or shelly feet may retain their front shoes until approximately one month before foaling, although the shoes must, naturally, be removed regularly and the feet pared before they are replaced. Attention to teeth is another important point and a veterinary surgeon should examine the mouth at least twice a year and rasp the teeth if it is found to be necessary.

Stabling at night is not essential and in many cases pony mares become restless if shut in, since it is their nature to live out night and day. The more 'breedy' types will appreciate warmth and comfort at night and there is also an advantage in stabling animals receiving a fairly large concentrate ration. In this last instance one feed can be given at night and another the following morning, the ration being equally divided between the two. This is a better arrangement from the digestive point of view as the horse's stomach is not designed to take large quantities of food at one sitting but rather to receive a little at frequent intervals. Horses should also have access to water, if it is not always available to them in the stable, *before* they feed and *never* after eating, as this could and frequently does, cause colic. If water is always in the stable it will be seen that horses will drink small quantities frequently, which does no harm; it is the large bucket or two buckets of water which are drunk immediately after eating a corn feed which does the damage.

With the increase in the daily intake of concentrate rations it becomes necessary for the feeds to be divided into two lest the stomach be overloaded. A good plan is to feed nuts for the morning feed and to give the bran, oats and whatever else is included in the evening. This feed can be given warm by mixing the bran with hot water before adding the other ingredients. One word of warning about stabling at night – it brings about extra work. Stables must be cleaned daily, except where the deep litter method of bedding is used, and so do not be over-anxious to stable your horse or pony at night. Remember, also, that it is important at all times that the mare gets a little gentle exercise each day, even though it is only for a short period when the weather is bad and makes conditions difficult. Dry cold, incidentally, does no harm at all and horses and ponies enjoy playing and rolling in the snow. The most unpleasant weather conditions for them are those combining cold and wet.

If boiled linseed is added to the diet twice weekly the bowels should remain in a good condition. However, should the droppings become hard at any time, immediate remedial action must be taken or the whole system becomes blocked. It is, in any case, advantageous to feed a bran mash once weekly, particularly after Christmas when there is very little moisture in the available grazing and mares are relying entirely on hay and hard food for their sustenance. Allow 2 lb. of broad bran for the smallest pony and up to 4 lb. for the larger animals, and make the mash as follows: take the required quantity of bran and place it in a bucket, adding to this two pints of boiling water to every 1 lb. of bran, a tablespoonful of common salt and one to three tablespoonfuls of Epsom or Glauber salts per mash. Stir the mixture thoroughly, sprinkle a small handful of dry bran over the top and cover with a sack, allowing it to 'cook' in the container for one and a half to two hours, after which time it may be turned out into a feed bowl and allowed to cool. Make sure that the water used for making the mash is boiling and make equally sure that you have thoroughly stirred it with your hand to satisfy yourself that it is not too hot before giving it to the horse. In cases where droppings have become very hard a twice weekly mash may be necessary for one or two weeks.

The aim is to produce droppings which are of a good colour, light to medium brown, and which break upon contact with the ground. Should an owner be faced with an animal whose droppings are much too loose, producing an almost liquid condition, food should be fed as dry as is possible for safety and veterinary advice sought. In the majority of cases neither of these extremes occur, but it is essential to recognize them if and when they do and to take the appropriate action.

Whatever the diet and feed constituents are, they will have to be changed gradually between the tenth month of pregnancy up to foaling time, since for ease of foaling it is better for the system to be relaxed in every way. To achieve relaxation it is sometimes necessary to reduce the quantities of nuts and oats and introduce greater quantities of moist bran, root vegetables, etc. In cases where mares are foaling in May, however, there may well be ample supplies of good grass during the last few weeks of pregnancy to produce the proper tone in the system, which in itself is a good reason for late April or early

May being the best time for the foal to be born. It is usually during this part of the year that the grass is most effective in producing a good milk supply for the expected foal.

I advised delaying the commencement of concentrate feeding for the smaller ponies at the beginning of winter and I would suggest that in the spring the quantity of food should be reduced somewhat earlier, certainly as soon as the new grass has started to come through. It is remarkable what good use these smaller, hardier types of animals make of the available food and they will certainly put on fat very quickly in the spring. Although it is not desirable that they should be thin it is certainly important that they should not be over-fat at the time of foaling.

In many cases these smaller mares are being used in an effort to produce larger offspring and while they themselves must not be fat their expected offspring must be given every possible chance to grow both before and after foaling. To this end it may, therefore, be necessary to increase the mare's food again once she has foaled.

As pregnancy advances it is quite probable that a change will be noticed in the temperament of the mare. Very often the quietest and most friendly animal becomes thoroughly crotchety as she nears her foaling time. There is no need to be alarmed about this, nor is it necessary to make excessive overtures to her in an effort to bring about a more happy frame of mind. She will return to normal again after foaling and be none the worse for her period of apparent ill temper.

Chapter 4

FOALING

Though individuals differ with regard to foaling signs
– some mares giving little or no indications of foaling,
even up to the last three or four days of pregnancy – the
majority start to show signs about one month before the
anticipated foaling date, which is approximately 340 days
after service (thus a mare covered on 15th May in one
year will be expected to foal on or about 20th April the
following year). Except on establishments where horse
breeding is a business, very few owners see their mares
foal, the majority of foals being born during the early hours
of the morning. Since most owners will not have a suitable
foaling box, as soon as weather conditions permit mares
should be left in the field night and day, although they
may still be brought to the stable for food and, in any
event, must be fed regularly and inspected at least twice
a day.

One of the earliest signs of foaling is the development
of the udder. At first this will hardly be visible but some
enlargement can usually be noticed three to four weeks
before foaling, particularly in the morning when a mare

has been stabled overnight. As time goes by the udder will increase until it is more obvious at all times, eventually becoming hard and shiny. During the last few days of pregnancy the dorsal muscles will become relaxed and quite obvious hollows will appear on each side of the spine, above the base of the tail. Between 24 and 48 hours before foaling a globule of wax will appear on the end of each teat. This may be very small and almost unnoticeable or it can be of considerable size and remain on the teat for a number of hours. The wax is a honey-coloured substance and ultimately drops off of its own accord, following which foaling can be expected at any time, although it is quite likely that a mare whose wax has dropped off in the afternoon will not foal until the early hours of the following morning. Once the wax has dropped, milk may begin to run and spots of this may be seen down the insides of the hind legs.

From now on the mare must be left in peace, although she should be looked at from time to time to make sure that all is well. She may stand for an hour in one part of the field or she may be seen grazing as normal; she may also be noted to take a quick walk up and down the field and then resume grazing. These are all signs of the early discomfort of foaling, the process of which begins when she gets down and the water bag appears.

Following the appearance of the water bag the two forefeet will emerge, one slightly in advance of the other. After this, the nose, which is resting on the forelegs, will appear, and from then on foaling will proceed. At this stage a mare will often get to her feet and lie down again several times, even when presentation is quite advanced, but again this is no cause for alarm. In the majority of mares the complete act of foaling is quickly over, the biggest effort being when the shoulders pass through the pelvic girdle. It is for this reason that one foreleg is positioned slightly in advance of the other, allowing the shoulders to be narrower than they would be if both forelegs were level. Once the shoulders are through the remainder of the body and the hind legs follow easily.

If you are in attendance do not disturb the mare during the birth process but let nature take its course, as at this time circulation is still continuing between the newborn foal and the as yet unexpelled afterbirth, technically known as the placenta, and it is extremely important to the well-being of the new-born foal. Shortly after delivery the foal will be seen to struggle and break the water bag, if this has not already taken place, and the first natural breaths will occur. At this time the mare will usually be lying quietly, recovering from the strain of giving birth. During the foal's struggles, or perhaps when, in due course, the mare rises, the umbilical cord will be ruptured.

At this time mares are particularly apprehensive and even the greatest old favourite may well become aggressive if her new-born foal is approached by a human. Go quietly to the mare's head with a feed or some other suitable encouragement in your hand, and attach a rope or rein to the headcollar. You are then in a position to control her should she become possessive.

Dealing with the navel cord is the next important step. The foal should be held on its side and cotton wool, freely impregnated with iodine, applied to the stump of the cord attached to it, after which the area should be dusted with an antibiotic powder. This dusting is important since germs can enter the new-born foal via an unprotected umbilical stump. Having given attention to this loose the mare, stand back and allow the foal to make its initial struggles to rise. These may be quite considerable and the most horrifying bumps will occur when the foal gets half up and then falls back. But do not be concerned. This is nature's way of keeping the foal warm, and allows it to get the use of its limbs, which at such an early age are incredibly shaky, by gradual stages. If the mare has placed herself near a fence action must be taken to see that the foal does not fall on to it in its efforts to rise.

Once the foal is on its feet it will not be long before it is thinking about its first meal. An indication of this will be when the foal starts to nuzzle the mare, in every place but the right one at first, but in due course it will approach the flanks and push its nose underneath. The colostrum, which is the name given to the first milk that a foal receives after birth, is particularly essential to the new baby as it contains many anti-bodies which help to provide resistance to disease. Round about this time the mare, too, will probably appreciate a warm feed, which may be made along the same lines as a bran mash but with water which is of a temperature suitable for the mare to be able to eat the feed immediately. There is no reason why a small portion of bruised oats and a little salt should not be added to the bran to make the meal more appetizing.

When the foal has fed it should have a bowel action. This may not take place immediately after feeding but it normally does so within an hour or two. This first dropping, called the meconium, is of a dark brown colour and ideally it will be the consistency of very thick cream. If a foal is seen to be constantly straining without any result it is possible that there may be a stoppage. Obviously this must not be allowed to continue for any length of time, since new-born foals are quickly weakened by such straining and the impaction worsens as it continues, and so it will be necessary to call in veterinary assistance without delay.

Although some mares cleanse quite quickly after foaling others may not do so for several hours. This need not cause immediate concern, but any mare that has not cleansed within four to six hours should be attended by a veterinary surgeon. No attempt must be made to remove the placenta by pulling.

If the foaling has taken place without your being present approach the mare, on finding her with her new-born foal, with confidence but also with some caution, and quietly take hold of her headcollar as soon as you can. Most mares are kindly disposed towards their owners but the protective instinct for their young is frequently uppermost at this time and they can be unpredictable. Having caught the mare, endeavour gently to take hold of the foal by placing one arm round its quarters and the other in front of its chest. It will undoubtedly struggle but it is essential to hold on, and it will soon quieten down, particularly if the mare is allowed to come and nuzzle it. Of prime importance is attention to the navel, as already described, although now this must be done with the foal standing and three people may be needed – one to hold the mare, who should have her head to the foal, one to hold the foal firmly and the third person to attend to the navel. This last person should not bend down too close to the foal's hind legs and should keep his or her head above the foal's body level. It is quite amazing how painful a kick from even so young a foal can be! So do keep near to the forelegs rather than to the hind.

Next, inspect the mare's udder to ensure that it is not still hard and shiny. If it is not it indicates that the foal has fed. The problem of knowing whether the first bowel motion has taken place is not quite so easy, but inspection under the tail may show traces of the dark, moist substance which has been passed. Sometimes it is impossible to know whether any meconium has been evacuated and therefore observation will be necessary for a few hours to see whether or not the foal is constantly straining. If there is no sign of the placenta hanging from the mare it may be assumed that she has cleansed successfully. For the first ten days after the birth of the foal a close watch must be kept on mother and child for signs of ill health. The former is the less likely of the two to give trouble but should she go off her feed and seem listless it is possible that a small portion of the placenta has remained and is setting up an infection and early veterinary attention will be required. Daily inspection of the foal's navel is a must, and any sign of a moist condition, which would indicate an infection of a serious nature, demands immediate professional treatment. It is also important during these first few days to ensure that both the foal and its dam have adequate rest, and it is advisable that they should be visited only by their owners and persons closely connected with them. It is great fun to show off the new baby to one's friends, but the temptation should be avoided for as long as possible.

An anti-tetanus injection should be given to the foal within four days of birth, and the course of injections for life immunity may begin when it is one month old. Do not be misled into thinking that tetanus is not prevalent in your area or because you cannot see a wound there is no possibility of infection. Tetanus lives in the soil and is ever ready to strike, in many cases doing so through small, unseen scratches. Owners who do not give their animals the available immunization, which is not expensive, must be considered negligent.

New-born foals need a great deal of rest and will frequently be seen lying down, but do not assume that a foal that lies down constantly is all right; make sure that from time to time it gets up and goes to the mare for a feed. Sometimes a foal will be seen to put its head under the mare but will not, in fact, be taking milk. Whether it is or not can be ascertained by inspecting the mare's udder which, if a foal is not feeding, will quickly become hard and shiny and cause her considerable discomfort. If such a state of affairs comes about it is more than likely that the foal is in some way not completely fit. Fifteen to twenty minutes a day spent just watching the mare and foal are a very good insurance against anything going wrong.

If a foal that is lying down is made to get up it will normally go to the mare and take a feed, after which it will probably have a canter round and take a lively interest in all that is going on, including the strange two-legged beast which invades its field from time to time. But beware of the foal that constantly goes to the mare, nuzzles the udder, goes to the other side and does the same, goes away and then returns and tries again, as this can indicate a shortage of milk, not a very common occurrence but one which it is not always easy to remedy. Be suspicious, also, of the foal that is listless and doesn't bother to feed; he could well be running a temperature.

Fortunately, problems such as I have mentioned are not common and most foals gain strength daily, although one common problem is that of scouring. This can occur in the foal when the dam comes into season at approximately nine days after giving birth, at which time there is an alteration in the milk. Usually this scouring is short-lived and causes little trouble, although it is very often necessary to wash the foal's bottom daily during this period to ensure that the skin does not become scalded by the scour and cause the hair to fall off. Two people will be required for this washing operation, one to hold the foal and the other to do the washing. The foal holder should take the

1. A foal ten minutes after birth.

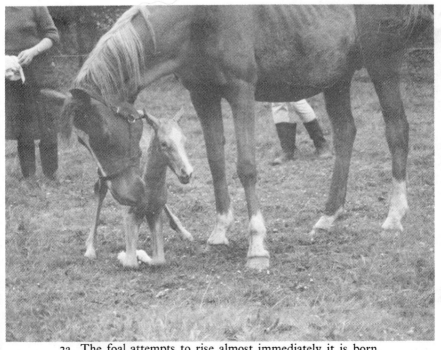

2a. The foal attempts to rise almost immediately it is born.

2b, 2c. This baby is just thirty minutes old.

3. By the time the foal is one week old it shows no hesitation in telling its mother that it is time for tea.

4. Mares and foals need plenty of rest.

5. A well-fitting headcollar is an essential for the brood mare. This one has the added refinement of a brass name-plate.

6. This charming foal is wearing a foal slip which has clearly seen much service. However, it fits his head properly and is not adjusted so tightly that it could cause him any discomfort.

7. The young foal should be held between the attendant's arms close up to the mare's left flank.

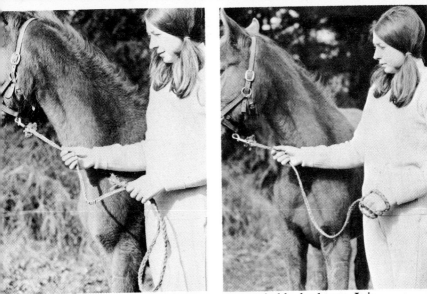

8a, 8b. The right and the wrong way to hold a lead rope. It is not advisable to wrap the rope round either fingers or hand at any time.

9. These two youngsters, both ten months old, have been taught to lead freely in either hand.

10. Snow has no harmful effects on young stock and there is no reason why they should not be turned out, even when there has been a heavy fall. They enjoy playing in the snow just as much as children do. These are both Anglo-Arab yearlings.

11. A good food container should be sufficiently large to allow both mare and foal to feed in comfort. This converted sink is also heavy enough not to be easily overturned, and has no sharp edges.

12. A young Palomino foal ready for the show ring.

13. A three-year-old part-bred Arab pony showing how well she has learnt her lessons. This is an excellent example of a horse in hand running out freely and straight.

foal slip, which will have been put on early in the foal's life, in the left hand, using the right to steady the fore-end of the foal. It is usually enough to place this hand on the shoulder and indulge in the usual wither scratching. The operation will be carried out with greater ease if the foal can be stood along a wall, particularly if the mare will stand quietly, perhaps eating her daily feed, while her offspring is being given attention. The person doing the washing needs to stand to one side of the quarters and always on the same side as the holder. A mild disinfectant and lukewarm water should be used for the cleaning and applied with a piece of cotton wool or gamgee tissue. When both sides of the tail and the tail itself have been washed quite clean any excess water must be wiped away and followed by a further drying-off with a very soft cloth. The task is completed by thoroughly powdering the damp areas and the tail with boracic or other suitable powder, which will help to dry the area quickly and prevent any soreness. Should the scour at any time during this period become foul smelling and of a light colour, veterinary advice should be sought. Foals may also scour when they start to drink water and it is extremely important to see that containers holding water are kept scrupulously clean.

Chapter 5

HANDLING AND TRAINING A
YOUNG FOAL

As I mentioned in the previous chapter, by nine days of age it will be necessary to handle the foal and to hold it, and so from the outset it is essential that it begins to learn obedience. To this end, if it is at all possible, mares should be brought to the stable daily, although obviously if all is well it is better not to subject the foal to this new experience for the first 24 hours of its life.

Teaching the foal to lead is of great importance in its training and paves the way to proper behaviour and obedience later on, but new-born foals are very often shy and difficult to catch in a field and so at first it is better to allow the foal to follow its mother into the stable. The mare can then be placed against a wall with an attendant holding her, and the foal encouraged to take up a position along the near-side of its mother. Once the foal is in the desired position approach it with arms partially out-stretched, placing the right arm around the quarters, slightly below the rump, and the left round the front of the foal's chest. Initially, the foal will almost certainly jump about, but one must hold on firmly and seek to calm it by speaking quietly.

Once a foal understands that you do not mean it any harm, but that you are determined not to let it go, the battle is beginning to go your way, although it is quite probable that for several days you will have the same kind of trouble. As time goes on, however, the period spent in struggling to escape will decrease, until finally the foal will be content to stand quietly. At all times keep the foal close to the mare's left side, that is, the right side of the foal against the mare's left flank. This will give it confidence since it will be touching its mother.

Once the foal has been secured and stands without making a fuss, lead the mare slowly round the box encouraging the foal to follow within your encircling arms. Push slightly from behind and be ready to restrain any sudden plunge forward with the left hand. After the first day or so a stable rubber may be used round the neck in place of the left arm. This will allow the attendant greater freedom of movement and also helps in teaching the foal to go forward from a slight pressure brought to bear behind its quarters. The stable rubber can be placed high or low on the neck depending on the foal's behaviour. At the outset these early lessons must be of short duration, but once the foal is calm and you feel able to maintain control with safety, the foal may be led back to the paddock and in time will be sufficiently quiet to be led both to and from the stable.

Each foal is an individual and temperaments vary

enormously from one to another, some being extremely nervous while others can be surprisingly precocious. Never at any time let go of any foal without getting it to understand that it has nothing to fear from you. All foals love to be scratched just in front of the wither, on the chest and at the base of the spine where it joins the tail. Therefore, once the foal has yielded to your demands to walk forward and follow the mare, scratch away at one of these spots before releasing the youngster completely – you will quickly find which is the spot your foal most likes to have scratched. If as you start the scratching you gently remove your restraining arms the foal will eventually stand to be scratched without having to be held.

When teaching the foal to move forward it is important that the hold on the neck should be released immediately before applying pressure on the quarters. If the foal is restrained, and so prevented from obeying your injunction to move forward, it may very well try to throw itself over backwards in its struggles to escape.

At the age of one week the foal slip, purchased some months ago and now oiled thoroughly, should be fitted on the foal's head, care being taken to see that it is adjusted neither too tight nor too loose round either the nose or the head. A slip that is too loose is dangerous because a hind foot may be trapped in it when the foal scratches at its head. A tight slip is almost as bad, as it will chafe the delicate skin. Not only is this uncomfortable for the foal but it may make it head-shy later in its life. Naturally the slip will have to be adjusted frequently to conform with the growth of the head. It should be possible to insert two fingers in the noseband of a well-fitted slip, so that the foal is able to feed without restriction, and the head strap should be sufficiently long to allow the noseband to lie comfortably below the cheek bones.

Fig. 4. The correct way to put on a foal slip. The youngster is positioned so that he can only go forward and therefore into the slip.

To put a slip on a foal requires a little tact and guile. Obviously, it is useless to approach the animal head on, holding out the slip in the hope that the foal will oblige by placing his nuzzle through the noseband for you. What the foal will do is to run back away from you, and you must position yourself so as to make it impossible for him to back away. With the foal placed like this any forward movement on its part must result in the muzzle being put into the nosepiece. To achieve this an assistant will be needed to keep the foal's quarters in position, or the single-handed owner must back the foal into a corner where it cannot get away. The former method is obviously the more satisfactory unless one is very skilled.

Once the foal is in position take the slip in the left hand holding it by the left cheek strap, having first buckled the noseband to the size which appears to correspond with the foal's muzzle. Approach the foal, speaking to it, and position yourself near its left shoulder, facing forward. The right hand should hold the right side of the slip, just above the junction of the cheek strap, so that the headband is turned slightly outwards, away from the foal's cheek; to hold the slip in this way the right hand travels below and behind the jaw bones. Gradually raise the slip until the noseband is as nearly as possible in line with the foal's nostrils, and then bring both hands gently backwards, passing the noseband over the muzzle. This action may well cause some confusion, since the foal will undoubtedly be frightened at feeling something strange upon its nose. Up till now it has not known any form of head pressure and it is natural for it to feel frightened when first restrained by the head. If you are successful in getting the noseband over the muzzle all that is now needed to bring the headband over and behind the ears is a slight flick of the right hand. The strap end can then be caught in the finger and thumb of the left hand, while the right is brought quickly back below the chin to fasten the buckle and complete the exercise.

If the foal struggles do not become alarmed, but make sure that your assistant is in full control of the quarters and is ready to push forward if the foal makes any attempt to stand on its hind legs. The one thing to avoid at this time is the foal throwing itself over backwards with fear. Buckle the head strap comfortably and while still retaining a very light pressure on the short strap attached to the ring at the base of the foal slip, scratch the foal on the wither and speak to it calmly. Endeavour if possible to persuade the foal to move a few strides forward. No doubt it will shake its head and express some annoyance, or even fright, but it is essential to gain the foal's confidence at this point, which you will best do by keeping calm. Give the command 'walk on' in a firm, quiet voice and reward the foal by slackening the tension with your left hand the minute it yields to your slight pull on the slip. Thus, it will associate moving forward when pressure is applied behind the ears, which is what happens when you take your first gentle, forward and downward pull on the lead strap. Encourage the foal by speaking calmly and ask for one or two more strides. If it struggles and runs backwards push its quarters away from you immediately; this will help to lessen the danger of its standing up on its hind legs and falling over. On no account at this early stage must you pull on the foal slip without having either

47

your right hand on the quarters or an assistant behind the foal to push it forward.

For the next few days lead the foal, exerting a gentle tension on the foal slip with the right hand applied on the quarters in the same way as when you were leading with a stable rubber. At all times keep the foal close to the mare's flank and foil any attempt at running back or standing up by either pushing forward from the quarters or moving the quarters slightly sideways, away from yourself. The main points to remember in these early leading lessons are the need for firmness and an insistence on obedience. Always reward immediately you have gained a response. Once the foal moves forward, even one or two strides, as a result of the slight tension on the lead strap combined with the encouraging push from behind coupled with the command 'walk on', immediately relax both your left and right hands. To begin with this relaxation will cause the foal to stop and once again you must repeat the commands to move forward. Very shortly the foal will connect the pull and push with the words 'walk on', and it will not be many days before you will only need to use the verbal order, backed up by the lightest of hand pressures. The importance of this early training is considerable and its value will be proved later on when you require the foal to move away from the mare and take a line of its own, rather than sticking close to its mother's flanks.

Most owners will not have the time to spend all or even half the day with their mare and foal, and therefore a routine must be established which fits your own personal time-table and gives the animals their reasonable share of care and attention. A small part of each day should, however, be devoted to the foal's schooling. Quite a useful plan, particularly when mares and foals are out all the time, is to visit them as early as possible in the morning to give the morning feed and at the same time to make your inspection of each animal. You should check that the foal is feeding and that neither he nor his mother have sustained any cuts or grazes, and that there is an adequate supply of water for the day. In the afternoon or evening, whichever is the more convenient, the mare and foal may be brought into a stable if one is available.

Such a programme allows one to give the foal its lead lesson and to handle it a little more thoroughly. As soon as the foal is leading well on the left side of the mare it must be taught to lead equally well on the opposite side of the mare, which entails reversing all the instructions previously given for leading. To begin with the foal will display a certain amount of resistance, as may the mare if she has never been taught to lead in this manner. However, you must persevere until both mare and foal go freely when led from the right side. It is not only a good exercise for all concerned, but is of particular advantage to the young animal as it encourages suppleness on both sides of the spine, which is an essential requirement for the later training under saddle. A great many young animals when old enough to be either broken by their owners or sent away to be broken professionally prove to be one-sided; that is, they tend to turn easily to the left, but have difficulty in bending to the right. The fault lies with the person who has had their early handling and who has allowed them always to be led in the right hand. Leading from both hands is also an essential control factor. Both at this early age and in later life the young animal should be taught to turn *inside* the person leading it. To turn in this way gives far greater control to the leader because it causes the animal to engage its hocks to make the turn and so reduces the likelihood of the horse swinging round and kicking out.

From as early an age as possible, certainly after the second week, introduce the foal to having your hands run all over its body. Start by placing your left hand on the left side of the neck and very gently running it down the foreleg, at first as far as the knee and then, as the foal accepts this handling, down to the fetlock joint. Do the same with the left hind leg and then reverse your position, applying the right hand to the right side of the neck and continuing as before. Young animals learn to tolerate such handling very quickly and in time you will be able to pick up each of the four feet, even if at first it is only for a second. As soon as the foal picks up a foot, and provided that the leg is not repeatedly snatched away, you can consider that you have made progress. As time goes by you will be able to hold each foot a little longer and this training will be invaluable when the farrier comes to dress the foal's feet. The feet will need attention after about three months and thereafter once a month.

As well as making every attempt to obtain obedience

from the foal, owners must also ensure that it does not become cheeky. This is particularly important with regard to colt foals, who in no time at all will become very precocious. Such behaviour, if unchecked, can lead not only to naughty habits but can be really dangerous for everyone concerned. Any attempt to nip, for instance, must be sharply discouraged, even if it means giving a light, sharp tap on the nose. Do not be amused by the foal that comes up to you, turns round, backs up and then kicks. This again is something that must be treated very severely if it is not to become a naughty and dangerous habit.

Lessons in leading must continue and be extended. The foal should be taught to stand still while the mare is led away from it, at first for a very short distance, and then the mare must stand while the foal is led away from her. These training sessions should be of very short duration and be given at least three times a week. On the other days of the week the foal may just be led to the stable in the normal manner, but always remember to change the side from which it is led on alternate days.

To vary the stable lessons, and as an alternative to hand rubbing, introduce a soft brush on the body and also on the mane and tail.

At no time during these training periods should a mare with foal at foot be left tied up, unless there is someone in attendance; to do so presents a considerable hazard to the foal, who may become entangled in the mare's rope.

Do not forget that if your foal is eligible for registration, a registration application form must be obtained from the appropriate society, duly completed and returned. If your foal has one grey parent and you are not quite sure of its own colour (for instance, if it appears to be chestnut but is not, to your eye, as chestnut as you have seen before) enter it on the form as being chestnut or grey. A grey horse is never born this colour, the ultimate grey hairs only appearing when the foal coat is shed. In some cases, however, the true colouring will not be revealed until the spring following birth.

Chapter 6

FEEDING, SHOWS AND MANAGEMENT

In all forms of animal life growing youngsters need good feeding, and the young equine is no exception. Of equal importance is the cleanliness of containers in which food and water are given. No matter how much, or how good, the food offered, a great deal of its value will be lost if feeding containers are dirty and make part of each feed unpalatable. If any part of the feed given is left in the container it must be removed and the manger or bowl thoroughly cleaned before being used for the next feed. If food is again left you can be satisfied that it is not because the container was dirty or the food stale, and must then decide whether the animal is leaving food because it is not well or because you are giving it too much. If both mare and foal seem lively, and if the foal feeds frequently from the mare, you are probably over-feeding. Try giving less food for several days and then gradually start to build up the quantity again, noting the length of time taken to clear up a feed and whether or not it is eaten enthusiastically.

Clean drinking containers are even more important,

since dirty and stagnant water will easily produce tummy upsets in the foal, resulting in a most unpleasant form of scour. A fast-running stream provides an excellent source of clean water, but stagnant ponds do not. If there is a water trough in the field make sure that it is kept clean and free from leaves, twigs etc. If an old bath or a similar container is used, endeavour to give it a complete scrub out once a week, particularly in hot weather. Feed containers, both in the field and the stable, should be cleaned daily.

If the mare and foal are to be fed in the field, a visit to a builders' yard may enable you to purchase an old-fashioned sink, a very easy container to clean and one that is neither easily knocked over nor has sharp edges upon which a foal can damage its legs. Wooden boxes are not suitable as feed containers as they are easily broken. Once they begin to break up nails and splinters become exposed, presenting a hazard to the mare and her offspring, not the least of which is the danger of one or other of them walking on a board through which protrudes the sharp end of a nail. Make sure that any feed container provided is large enough for mare and foal to feed together in comfort. If at all possible avoid leaving buckets or other handled containers in the field with the mare and foal. The latter is sure to kick a bucket over and may severely frighten itself by catching a foot in the handle. Even if it does not hurt itself by doing so it may be frightened into running off and damaging itself against a fence.

From the age of about two weeks foals start to feed independently of the mare, although at first this may amount to no more than a nibble at a few oats from the mare's daily ration and the tentative sampling of a little grass. Every endeavour should be made to encourage the foal to feed on his own as it will accelerate his growth rate, and later on, when weaning time comes, the foal will more readily accept the food provided for it and will suffer less of a setback. When the foal is about a month old it will be necessary to make a distinct increase in the rations provided for the mare so that there is enough for the foal to take its share.

As a rough guide, a foal will consume daily approximately 1 lb. of concentrate food for every month of its age, so that by the time it is weaned it will probably be taking about 4 lb. per day. With the exception of the larger nuts, which they do not find easy to eat in extreme youth, foals will enjoy the same kind of food as their mothers. All food should be given in appropriate containers and the mare and foal left to feed in peace. A constant return to the stable while animals are feeding is very much on a par with a person walking into your dining room at lunch time, ruffling your hair each time they pass your chair – a habit which I think none of us would appreciate, even from our greatest friends.

Although the foal should be encouraged to eat as much as is reasonably possible only give food at the regular meal times. Do not fall into the habit of feeding titbits – it is *not* a good idea. A foal that receives titbits soon learns to nibble at you every time you appear in the hope that he is going to get something. This nibbling, though amusing in the early stages, is far from funny when the animal is a two- or three-year-old. Sometimes, indeed, horses or ponies who have been given titbits in their early youth resent anyone entering the paddock or stable without such an offering; even, in some cases, going so far as to lose their tempers, either biting, or turning round and kicking, if the anticipated titbit is not forthcoming. These are not pleasant habits and they can be dangerous, particularly where children are concerned.

If the mare and foal can be brought once a day to the stable it is helpful, from about the second month of the foal's age, to spend time in persuading the foal to eat from a separate container, so that one knows just how much food it is taking. Separate feeding at this stage is also good preparation for the time when the foal will have to feed by itself after weaning. Bring both mare and foal to the stable, having previously mixed whatever feed is given and put a portion of it into a plastic bowl, or some such other convenient receptacle. Allow the mare to feed from the manger and try to persuade the foal to take at least part of its feed from the bowl which you yourself are holding. You will find it easier to stand fairly close to the mare's head rather than in an opposite corner, since the foal, until now, has been used to feeding in close proximity to its mother and from the same container.

To begin with you may find that you have to wait quite a long time before the foal will eat even a mouthful, since it will be occupied in trying to feed with the mare as usual. It won't be too long, however, before the foal realizes that food, even if it is from a different bowl, tastes just as good. Needless to say, in the early stages the mare will probably finish her feed first and on quite a number of occasions the foal will refuse to clear up. Do not be worried by this; in due course the problem will resolve itself, particularly as the foal grows and requires more food.

One of the most important items connected with the young horse or pony is the control of worms, so don't bury your head in the sand and say 'my foal does not have worms' because, unless proper worm control is practised, it most certainly will have. The fact that the mare and her foal are in good condition is no indication that the foal is not carrying a sufficient number of these parasites to cause damage sooner or later. A proper worm control policy should be decided upon and carried out, not only with the mare, but the foal as well. There are excellent preparations available for the control of worms and, if your paddock is small, the daily removal of droppings will help considerably in controlling re-infestation.

The two main types of worm which infest the horse are Strongyle, or red worms, not often visible to the naked eye but capable of causing death if not controlled, and Ascarids, or white worms. These latter can be quite large and are in the majority of cases found only in the young animal up to the age of three years. Reference was made in an earlier chapter concerning worm control and no apology is made for doing so again; *worms can and do kill*. If you are in any doubt at all send a dropping sample, collected from separate motions over a period of

12 hours, from the mare and the foal to your veterinary surgeon. He will then make a worm count and advise you should any treatment be necessary. In all cases it is advisable to worm foals thoroughly, by whatever method is chosen, at three months of age and again soon after weaning.

Up to the time of weaning it is obviously not easy to ensure that the foal receives the correct amount of worm dose if it is being fed with the mare, which is another reason for teaching the foal to feed from its own bowl.

Should your local show offer a suitable class you may well be tempted to exhibit your mare and foal, and to do so once or twice during the season does no harm. Such an outing does, however, place a considerable strain on the foal, and should not be a frequent occurrence. Mares with foals at foot are normally exhibited with manes plaited and tails either plaited or pulled. Foals, on the other hand, are not often plaited for show, particularly when they are very young, although they may be plaited later in the season when they have almost reached the time for weaning and the mane has had a chance to grow sufficiently long. The foal, by the time you are ready to show it, will undoubtedly have learnt to be led properly, and should be quite used to being taken away from the mare to walk up and down a field, and to standing still while the mare does likewise.

One of the biggest problems connected with showing a mare and foal is that of loading them into the transport which is to take them to the showground. The matter is simplified if owners have their own trailers or horse-boxes, as they will have time to practise loading and unloading, but even if this is the case it is still important to go about the job, whether for practice or for real, in the manner which is likely to be the most successful. Do not attempt to lead the mare into the box leaving the foal to follow loose or to be led in by another person. It is highly probable that even if the mare goes in the foal will not do so. When this happens the mare, in concern for her offspring, will immediately try to get out, and quite frequently will do so. This sets a bad example to the foal, who is not only frightened itself but probably thinks that its mother must be frightened as well or she wouldn't have tried to escape. The whole operation is made far easier if the foal is loaded *first*, which may sound strange, but which is not as illogical as it seems since there are very few mares who will leave their foals. One does not, therefore, need to bother about the mother – she will automatically follow her baby.

One must first, nevertheless, load the foal, and two people will be needed to do this. They should take up positions on either side of the youngster, and then join hands just below the foal's tail and in front of its chest. The foal is then 'cradled' between the two pairs of arms and may be easily propelled up the ramp of the box or trailer. Do not bother about the mare if there is not a third person to lead her; either of the helpers handling the foal can just take hold of her lead rope, in the same hand that is behind the foal's quarters, and she will come quite willingly behind the foal. The lesson to be learned from this

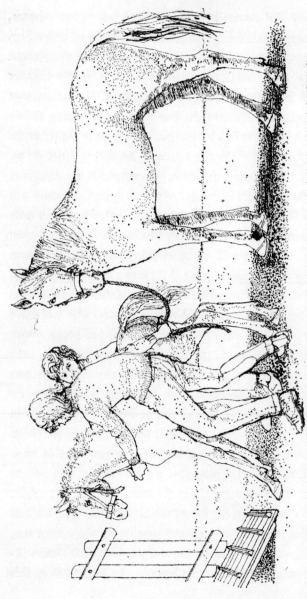

Fig. 5. Loading a foal into a trailer. The mare, whose lead rope is held in the right hand of the nearest attendant, will follow the foal into the box rather than be separated from it.

exercise is that the foal, although somewhat startled at being so to speak 'manhandled' into the vehicle, will quickly become at ease when it discovers that its mother is close on its heels and remains near at hand. The attendants handling the foal can make much of it and if the occasion is just a practice run it can be given a feed; in fact to feed both mare and foal in the box, even on the morning of the show, will help a great deal to avoid trouble when it is next necessary to load them.

Plenty of time must be allowed for loading, for giving the feed, travelling to the show and allowing the animals to rest on arrival. The foal is bound to be frightened and tired by the journey, and at least an extra hour should be allowed so that it and its dam may stand quietly in the vehicle before being unloaded and introduced to the excitement of the showground. Both will benefit from the rest and will come into the ring fresh and alert. Mares and foals should be allowed to travel loose in whatever form of transport is used, and they do not need to have their legs bandaged. A tail bandage may be used on the mare, but it is quite likely that the foal, during the time that it is standing in the box at the show, will make every effort to remove this new found plaything. If it manages to accomplish its object the tail bandage will, at best, get stamped around the dirty floor or, at worst, chewed into little pieces!

Be sure to offer both mare and foal a drink while on the show ground, but if they are to be left in the horse-box or trailer for any length of time without an attendant do not leave a hay-net tied in the vehicle, in case the foal paws at it and gets entangled in the mesh. Mare and foal, after being offered a drink, may of course have their usual feeds, but it is preferable that they should be given after the class rather than before it.

Mares may be shown in snaffle or double bridles which must, of course, be spotlessly clean, while foals are shown either in leather or white web foal slips to which are attached white leading reins. The feet of both mare and foal should have been recently trimmed and can be oiled before entering the ring.

The owners of pony mares may well be anxious to allow their children some riding during the summer holidays following the foal's birth. Assuming that the foal has been born in April or May, three months from the last of these dates is well within the latter part of the school holidays and, while not ideal, it is possible for mares to give a little pleasure to their young owners and continue to rear their foals. But extreme care must be taken both in the management of the pony and of the foal. In the case of the former, it is important that when first starting to ride her care is taken to see that she is only worked for very short periods and at slow paces. Some mares become very agitated when asked to leave their foals and the most docile can become a little difficult to handle, but the majority quickly realize that within a short space of time they return to their offspring and they will settle into a routine without too much trouble. The work must be limited to the slower paces both from the point of view of the mare's condition, which will be very soft, and also for the well-being of the milk that the foal will take on the mare's return. Fast work, causing the mare to sweat, has an effect upon the whole system and thus upon the milk as well. Milk ingested by the foal in this state quickly gives rise to an upset tummy.

Because of the mare's soft condition she will be as susceptible to galling as she was when she returned from stud, and care must be taken to see that she is not injured in this way.

While the mare is being ridden it is vital that the foal should be shut in a box from which it cannot escape or even make an attempt to escape. It is these attempted escapes, such as when a foal tries to jump through an unguarded window, or over the bottom half of a stable door, which can frequently end in tragedy. If the foal has been properly handled and has learnt to walk away from the mare and to stand while the mare is walked away from it, the foundation will have already been prepared for the mare to be taken away for a ride. Similarly, if the foal has learnt to eat separately from the mare, the placing of the usual feed container in the stable will encourage it to eat a feed while the mare is out.

The first occasion on which mare and foal are parted must be very short, a quarter of an hour being sufficient, or even less if the mare becomes extremely distressed and begins to sweat excessively. This period can gradually be increased until an hour away from the foal is quite possible, but owners must realize that the mares cannot be taken away for use in gymkhanas, Pony Club events, paper-chases and the like. The mare's primary task at this time is to rear her foal, and any riding that is done must be subservient to this occupation. If after a few days the mare does not settle and it is found that the foal starts to scour or is otherwise affected, riding must be given up until after weaning time.

Chapter 7

WEANING

Weaning may take place, subject to the foal being well grown, at any time after four and a half months of age, although in the case of precocious colts, who are already becoming a nuisance, it is wise to consider having them gelded whilst they are still on the mare. This can be done when they are about four months and they can be weaned a month afterwards. The gelding of colt foals while they are on the mare is no hardship and very little setback is sustained, although it is, of course, necessary to consider each colt as an individual and to obtain veterinary advice concerning the most propitious time for the operation to take place.

From the time it is weaned until it becomes a yearling is a very important time in the foal's life. The parting of mare and foal means that the latter must, from then on, continue as a separate entity without the benefit of its mother's milk. It is therefore essential that foals should not only have been well-fed while on the mare, but should also have been introduced to all the main basic foods, so that once the upset of weaning is over they will quickly settle down to eating again. If a mare has been ridden daily for short periods, or even once or twice a week, the foal will already have learned to accept being parted from its mother during these periods, but by the time it is five months old the parting should be complete and permanent as far as the mother-and-child relationship is concerned. Certainly it should not be delayed after the foal has reached six months. Should there be in the district another person whose mare has had a foal in the current year, it is a good idea to combine forces, so that the two foals may be weaned together. The object in this is twofold; first it can be arranged for the foals to live together in a box on one establishment, and secondly it might be possible for the two mares to live on the other one, out of hearing of their foals. There will be much calling at this time and it is far less disturbing both to mares and to their offspring if they cannot hear each other. Such an ideal arrangement as I have suggested is not, of course, always possible, and it will then be necessary for the mare and foal to be parted singly. Do, however, try to send the mare away, if only for a month or six weeks. Even after as long a period as this it would still be possible for the mother-and-child relationship to be revived if they were turned out together, and with it would come all the problems of weaning once again. If possible it is better to leave the mares in different surroundings, or alternatively send the foals away, until after Christmas. An advertisement in a suitable column of a local paper may well produce another owner with a similar problem and is well worth the small cost involved.

At weaning time one is faced with two separate problems: the first to get the foal to settle down and feed well, the other to dry the mare off. If two foals are to be weaned together the companionship nearly always helps the youngsters to settle down without too much trouble. They will, of course, call and might even attempt to escape, so be sure to close the top door of the box and make certain that the window is sufficiently protected, by either a grille or a wire mesh, so that they cannot get their feet through and injure themselves. The box should be deeply bedded and hay left ready for the foals. Hay is best fed from the ground at weaning time to eliminate the danger of the foals becoming entangled in a net. Once they have settled down, however, feeding hay in a net, tied sufficiently high to be out of reach of pawing feet, is the most economical method and one which allows the owner to know how much is being eaten.

If two foals who have not been previously introduced are to be weaned together, there will be a certain amount of fighting. But, providing that both are weaned on the same day, what fighting there is will not amount to anything serious, since the young animals will be much too concerned about their mothers, and will shortly find solace in each other's companionship. During the early part of the weaning period it is essential to harden one's heart and not allow oneself to become over-sentimental about the foals' distress at losing their mothers. The youngsters will probably eat and drink very little for a few days, and in some cases will rush round the box getting themselves into a sweat. Do not be upset by this behaviour, for it is quite natural and inevitable. Once the foal, or foals, discover that there is no answering whinny and that food is always available to them they will quickly enough decide to take a lively interest in the latter.

When entering and leaving the box in these early days take care not to let a foal push past you and escape, for if it does it will run wildly round and will probably, in its blind anxiety to find its mother, rush headlong into the nearest available wall or fence.

It will probably be necessary, if you have only one loose box, to leave the cleaning out of the box until the foals have settled down, and so fresh straw will need to be added daily as necessary. At the end of a week, most foals have become resigned to this start to their grown-up life and may be led out for exercise. Owners will have to decide for themselves how soon the foals may be turned loose in the paddock. Much depends on the height and nature of the fencing and also whether it has been possible to send the mares away a sufficient distance for them not to be within call of the foals.

If only one foal is to be weaned a quiet elderly companion will be needed and often a gelding, or a donkey, will make a very good 'nursemaid'. But this older animal must not, of course, be wearing hind shoes. When the day for turning out the foal or foals comes, a little bit of common sense will be required. The foals will not have had any grass for a week or ten days and they will undoubtedly be anxious to graze; this desire for grazing will be helpful to us and can be encouraged by reducing the feed on the evening before turning out and not giving any feed at all the following morning; hay, of course, should be available throughout the period.

Lead whatever animal is used as a companion into the centre of the field, and then bring out the foal or foals.

Take them as near as is safe to their new friends and then turn them loose, ensuring when you do so that the heads of the young animals are towards you; many foals on being turned out for the first time after weaning will kick from sheer excitement. Foals will also gallop round the field and on coming close to humans will kick out in play. These high spirits are quite natural but it is wise to keep a wary eye on the foals and their antics since kicks of this nature can be extremely painful for the recipient, to say nothing of being dangerous.

For the first quarter to half an hour, retain your hold on the 'nursemaid' and let the foals gallop round even if they give you more than a few horrifying moments by galloping towards fences and looking as though they cannot possibly stop in time to prevent crashing into them. On most occasions they do stop and usually they will at last return to the animal standing in the middle of the field. Once this initial excitement is finished their desire to graze will get the better of them, and in most cases it is not long before they settle down. They can be further encouraged to remain in or near the centre of the field, away from the dangers of fences and corners, if their companion is given a feed, since he or she will then remain in the same spot for some time after you have left the paddock.

We must not, however, neglect the mare at this busy time as she will also need attention, particularly to the udder. The sudden parting of the mare from her foal will mean that she is still carrying milk and while the udder itself will need certain treatment we must regulate the mare's diet and management towards reducing the flow of milk.

If the mare is stabled at weaning time, until such time as she has dried off all concentrate food should be curtailed and the water restricted to one bucketful twice a day. Hay should be available to her at all times and each day after her evening drink she should have a bran mash to keep her bowels in order and help the whole system to adjust. If she is quiet enough to leave in a well-fenced field she will not need the bran mash, as the grass, unless it is very dry, will act as a laxative. One must, however, keep a close watch on the droppings and if they become dark and hard a daily mash will have to be given until they return to a normal consistency.

The udder should be eased out (or milked) twice a day, but do not make the mistake of milking the mare out completely, since this will only cause her to produce more milk. Ease away enough of the milk to leave the udder slack, and as soon as possible reduce the milking to once a day. This may be done whenever it is found that a mare who has been milked out in the morning still has a slack udder in the evening. When once-a-day milking has been established over several days, and if the udder seems reasonably comfortable at milking time, it should be possible, by the end of the second week, to reduce the milking to every other day, when most mares will have dried off sufficiently not to be in discomfort or danger. If during this time the udder should become very hot and swollen, and particularly if the swelling is more pronounced on one side than the other, your veterinary surgeon should be consulted in case the mare has mastitis, a condition which needs expert professional treatment.

Just as the foal will call for its mother at weaning time so the mare will call for her foal, and it is for this reason that they are best out of earshot of each other. Mares can be as foolish as foals at this time, often attempting to get out of stables and paddocks from which they have previously made no attempt to escape. It is, therefore, just as important to confine the mare in a safe stable or, if she is quiet, in a suitable paddock, as it is to confine the foal.

Once the mare has dried off she may return to her normal life; being ridden or just turned out to grass with suitable feeding throughout the coming winter. It will probably be necessary to have her teeth rasped and in any case the usual monthly attention should be given to her feet and shoes. Attention to the matter of worm control must also be continued, since by the following summer your paddock, which has normally been keeping one animal, will have to keep two and if worm control is not practised the land will quickly become infested.

At, or soon after, the weaning of the foal, it is advisable to powder both mare and foal thoroughly with louse powder. Lice are nothing to be ashamed of and are frequently found on horses and ponies. Sometimes these parasites come from other animals, but more often it is quite impossible to determine from whence the infestation has come. But lice, like worms, are easily controlled. There are many good brands of louse powder on the market and instructions for their application are to be found on the containers. Sometimes it is impossible to see the lice on the animals at all, and for this reason it is a wise policy to powder regularly, certainly at the beginning of winter and again in the spring. Lice can and do bring about poor condition in horses and ponies and also cause considerable irritation and discomfort to their hosts.

If a mare has been wearing a headcollar throughout the winter, it can be removed after weaning for a few days so that it can be cleaned and re-oiled. This simple treatment will make it supple and pleasant for the mare to wear during the winter months and will also give protection to the leather. It is also advisable to ensure that no buckles or their tongues need renewing and that none of the leather parts needs stitching. Headcollars are expensive to buy and there is nothing more annoying than losing them in the field through want of a saddler's attention.

Up to the age of one year is a particularly vital period in the life of the young equine. It must be well fed, housed at night if possible and also taught a vast number of things which will enable it to be happy and manageable in later life. For the first few days after weaning the foal will undoubtedly be restless and will probably not be anxious to eat its food. Reduce the concentrate ration by half and divide it into two feeds. The reduction of food for a week or so will do no harm and may well prevent the foal from going off its feed entirely. The importance of clean water and feed containers has already been stressed and is no less important now. Weaning foals must be given every encouragement to eat well and it should be a daily routine to wipe out the manger and water bucket, and to give both a good scrubbing once a week. If rock salt is kept in the manger it will be found to sweat in damp weather and will make the feed too salty to be palatable. A good plan is to remove the salt and wipe the manger before giving each feed. If water buckets are left in the stable these should be suspended in a bucket ring at some distance from the ground; foals when playing about will be sure to upset the bucket, which will not only make the bed wet and uncomfortable, but obviously also means that they are left with nothing to drink.

Food, during the winter months, and indeed always, must be of the best quality obtainable and the concentrates as fresh as possible. Two owners may care to combine when buying such things as bran, oats, nuts, powdered milk, flaked maize, steamed barley, etc., and to divide the quantities between them. Presuming, for instance, that there are two foals to be fed and that they take 4 lb. of oats each per day, 1 cwt. of oats will last two weeks. Whereas, if one owner purchases 1 cwt. of oats to feed one foal the last part of this amount will be almost a month old before it is finished. Once oats are bruised or crushed they tend to keep less well than when they are whole, and bran and nuts are also capable of deterioration. All food should be kept in proper containers, which must have lids as a protection against rats and mice and which will prevent it from being affected by the moist atmosphere. If you do not have a proper corn bin, large plastic dustbins make excellent food containers.

Assuming that a foal is being weaned in September, and the weather during the autumn is good, there is no reason why it should not remain out day and night until mid-October or the beginning of November. Generally, two feeds a day should be given, although during the first month or six weeks after weaning this is not essential. It is in fact possible that foals will not eat two feeds a day if there is plenty of grass, even though the best of its feeding value has gone.

Chapter 8

FEEDING AND TRAINING
THE YEARLING

From the beginning of November foals should, if possible, be housed at night and their feeding properly planned to promote good growth. General management of the foal from this time is directed at obtaining obedience and sensible behaviour.

Since the foal will already have been introduced to the main feed ingredients there should be little difficulty in getting it onto a good basic ration. The same guide to quantity may be used as before, that is, 1 lb. concentrate feed for each month of a foal's age, up to maximum of 5 lb. or 6 lb. per day. The smaller animals will probably not take so large a quantity, but the bigger hunter type foals may well need more. Owners must be prepared to experiment to arrive at the correct ration for each individual. As always, the quality of the food is of prime importance and everything should be clean and fresh.

Once the foal is stabled at night, a routine must be formulated, particularly in regard to the regularity of feeding times. Whilst some horses are undoubtedly brighter than others, as a race they are not highly intelligent;

they are, however, creatures of habit and are more likely to do well if a daily routine is followed without variation. For instance, the animal who is never sure at what time his next feed will arrive quickly becomes unsettled and restless.

The principal foods to which the foal will be accustomed already will probably comprise horse and pony nuts, bran, bruised oats, sliced apples and root vegetables and possibly boiled linseed, but there are many additional items which may be added and each owner will have his or her own special favourite. One excellent preparation is powdered skimmed milk. It is obtainable from most forage merchants, but usually in $\frac{1}{2}$ cwt. bags, so it is sensible to try and find one or two others who will be prepared to share a bag with you. This milk powder may be added to the daily feed at the rate of 2 oz. to 8 oz., depending on the size of the foal. Larger quantities can be fed but are not usually necessary. As the powder is very fine it must be carefully mixed. The best way of ensuring that it is spread evenly through the feed is to place the usual amount of bran in a feed container, slightly damp it, stir well and then sprinkle the powdered milk over it, mixing thoroughly before adding any other ingredients. Flaked maize is another useful food, particularly in cold weather, but as it is heating it must not be given in large quantities, a $\frac{1}{2}$ lb. daily being sufficient for a foal likely to mature to 14 h.h. Smaller, pony type foals will take proportionately less; hunter foals may be given up to 1 lb. daily. Maize should be discontinued as soon as milder weather comes;

it is normally most useful from December until the end of March. Rolled, steamed barley may also be used, but, again, in comparatively small quantities and only during the colder weather. Quantities fed may be up to 50 per cent more per day than that suggested for flaked maize, but neither of these foods should be given to ponies with any tendency to become over-fat, since they encourage over-heating of the system, which may result in laminitis.

Cod liver oil is a necessary additive for the promotion of growth and bone development, but some animals do not take to it easily and it must, therefore, be introduced into the feed gradually or given in the more palatable cake form, known as Codolettes.

There is, in fact, a bewildering selection of mineral and vitamin additives on the market. Most of them are excellent products and can be of considerable value under certain circumstances; they are not, however, designed to be fed indiscriminately and owners should seek veterinary advice before embarking upon their use.

A suggested daily programme for the weaning foal might be as follows:

In the morning. Feed, after which the foal should be turned out and, if time permits, the box cleaned and made ready for the youngster's return.

In the afternoon. Bring the foal in, handle and educate as required, and follow with afternoon feed. When the hours of daylight are short, and the afternoon feed is early, a final check-up should be made later in the day to make sure that all is well and to top up the hay-net and water bucket.

The morning feed should consist of whatever ration of horse and pony nuts is given. If, however, nuts are not used this first feed should be composed of half the daily ration of other concentrates. Having fed the foal, probably before daylight in the short winter days, an inspection of the manger after breakfast will ensure that the feed has been cleared up, and the young animal may then be turned out.

Turning out is not just a case of taking the animal to the field, removing the rope from the foal slip and turning it loose. It must be taught to leave the stable in an orderly manner, to walk freely in hand and to stand quietly inside the field until it is released. When leading horses and ponies, particularly young ones, *never* wrap the rope or rein round your fingers, but double it in the palm of your hand. The former method carries the risk of badly torn skin and possibly broken bones if the animal pulls away and the rope tightens. The doubled lead rope gives just as much control and is far safer. Should the morning be frosty any ice on the water trough must be broken sufficiently to allow for drinking. Unless the weather is extremely bad there is no reason to bring the youngster to the stable again until half past two to three in the afternoon on the shortest days, gradually extending the period outside as the days lengthen. It is essential that the young horse should have daily exercise, and there are very few days throughout the winter when it is not possible to get them out for at least an hour or two; the few days when it is impossible to turn out young stock being those when there is extremely cold rain, or perhaps a blizzard. The fact that there is snow on the ground need not deter you from allowing the foals to run out for they will thoroughly enjoy romping about in these conditions.

For the first few weeks after weaning do not attempt to clean the stable while the foal is in it. This chore can be done while the foal is out in the field, and the stable will then be ready for its return later in the day. As soon as the stable has been cleaned water bowls, or buckets, and mangers can be scrubbed, and a filled hay-net hung in its usual place.

Leading out and in is a continuation of the foal's early training. Remember, however, to lead from alternate sides on alternate days, so that the foal never becomes accustomed, both mentally and physically, to being led from one side only. It is also a wise plan to make the foal stand still, with its legs well placed and on a slack rein, before allowing it to enter the stable in the evening. This is valuable training if the youngster is destined for the show ring, since it teaches the foal to take up a good, balanced position, automatically, when asked to halt. Once the foal is in the stable, spend a short time handling it and picking up each leg. Do not attempt to tie the foal up and leave it, but slip the rope through a ring placed high in the wall, holding the slack end in your hand, while you are handling the body and legs, etc. Should the foal pull backwards you will then be able to play out a certain amount of the rope and avoid an accident. In time, of course, the foal must realize that the rope, when attached to the ring, means that it must stand still.

When the foal has submitted happily to the handling exercises it may be released and the afternoon feed can be fetched. This feed will consist of bran, oats, maize or steamed barley, sliced root vegetables or apples, milk powder and/or cod liver oil if used. The quantity will be governed by the size of the foal and also by the amount of the nut ration which it received in the morning.

Assuming that a foal takes 6 lb. of food a day, half of this may be fed as nuts in the morning, and the other half of the mixture in the evening. Never allow the youngster to rush you for its food, but make it stand back while you tip the feed into a manger or place the food bowl on the ground. Hay, of course, must always be available and should be of a variety which does not contain too many hard stalks. A good meadow hay is excellent for the delicate mouth of a foal. This soft hay may be gradually discontinued as the youngster grows older and is more able to deal with the harder types of hay. Remember always to leave a salt lick in the stable or in the field.

Very often foals who have been feeding quite well will, for some reason or another, go off their food; perhaps not clearing up one or other of the feeds, or leaving some of both. This does not necessarily mean that it is being over-fed, but it is important that the ration should be reduced as quickly as possible, before the youngster completely goes off all forms of short feed. It is far better to reduce the diet by half, or even more, for a number of days until such time as the youngster is clearing its feed up readily, and then gradually to increase it again, rather than to allow the youngster to lose all interest in concentrate feed for several weeks – something that can, and quite often does, happen. The result, of course, is that the youngster quickly loses condition and a set-back in development occurs. Like children, foals sometimes become tired of their food and a great deal of ingenuity is required on the part of the feeder to tempt them back on to it again. In these cases such delicacies as sliced apples and carrots will often tempt the shy feeder. Should one experience food rejection by a foal the immediate, commonsense step to be taken is to check that the food it is getting is fresh and palatable. Food is easily contaminated and, indeed, its palatability can be affected by its being stored near to disinfectants. The taste of such substances quickly spreads into the food and it is then not surprising that feed tins lose their appeal.

Good feeding is obviously important, but one must not be tempted into going to extremes and over-feeding young stock. If the foal has to remain in the stable for several days, through injury, perhaps, or extremely bad weather conditions, the amount of concentrate food must be cut, particularly the oat ration, or the youngster will become too lively and possibly uncontrollable when you next wish to take it to the field. When the concentrate ration is cut, however, hay must be increased accordingly, since there will be no grazing period during the day. At such times and for the same reason bran mashes must be introduced to keep the bowels in order. As with adult horses, the youngster's droppings should break easily on contact with the ground, and should they become hard and very dark in colour the matter must be remedied immediately by feeding laxative foods. A similar cut in the concentrate ration, particularly in the oat content, may be made if a foal, even though taking its normal daily exercise, becomes too lively. Nothing is worse than a silly youngster whose antics are a danger to himself and his attendants and who through high living tends to get into bad habits. None the less, food must not be cut to such an extent that condition is lost. It is better to substitute a less heady food for the ration of oats, and it is here that properly balanced nuts can be helpful.

Care must also be taken to see that young horses do not develop any stable vices, such as chewing the woodwork. Many of them will try, and the cure is to paint the woodwork with non-poisonous creosote or some other deterrent, but *not* with any paint containing lead. Youngsters will often nibble the ropes by which they are tied and so these too must be painted with something unpalatable. Creosote is rather too messy, but soaking the ropes in a solution of bitter aloes can be very effective.

From about the end of January, as the days become longer, foals, or as they are called after the turn of the year, yearlings, will have longer periods out of doors. This is all to the good, but until such time as they are turned out entirely, probably towards the end of April, the strict daily routine should be maintained. Once the yearling has learnt to stand quietly when tied up its stable education can be carried a stage further and it can be taught to move over from one side to another in the box – a useful accomplishment for any horse as it allows one to clean out the box without going to the trouble of moving him out of it to do so. Care must be taken, however, and particularly with a young animal, that the cleaning implements – forks, brooms and so on – are not left lying around for him to fall over. In fact, it is a good rule never to allow any implement in the box except the one you are using. Move quietly round the young animal but make it constantly aware of your whereabouts, remembering always that the use of the hind legs in time of fear is the horse's natural defence.

To teach a young animal to move over in the box, place yourself so as to be out of harm's way but in a position from which you can obtain a response to your command. It is almost impossible for the horse to have its head and its hind legs facing in the same direction, so when you are teaching the yearling to move across the box, quarters going from left to right, place yourself by his left shoulder, take hold of the foal slip in your left hand, draw the head towards you and at the same time place your right hand on the quarters and push them away from you, giving the verbal command 'over' as you do so. Constant repetition of this instruction to both directions (the aids will be reversed when you want the quarters to move from right to left) will very quickly teach the youngster what is wanted. Aim always to discard the use of the pushing hand as soon as possible, and always preface the command 'over' with some spoken introduction, which will act as a warning or preparation for the command which follows. As response to the lightest of pressure on the quarters, combined with the command 'over', is gained, so the hand on the foal slip may be released. It will not be long before you are able to stand on the left or right, behind the quarters, at a prudent distance, and merely give the command 'over' to obtain the required result. Always remember to place yourself well to the left if you wish the animal to move right and vice versa. If you stand immediately behind the horse saying 'over', it won't work. The animal, not knowing which way you want it to go, will only become confused. Once, however, the animal has moved in response to your order, reward him with a calm word of encouragement. Animals as a whole respond not so much to words but to voice inflection, brisk sharp words resulting in brisk sharp movements and gentle, calming words having the opposite, quietening effect.

The winter care of weanling foals has been described on the assumption that stable accommodation is available at night, and obviously it is an advantage if foals can be housed in the way described. On the other hand, stabling is not essential to the foals' well-being, although if they are left out similar programmes of feeding and management will still be necessary. If it is at all possible, however, bring the animal in once a day for feeding and handling, even if you have only the use of a simple field shelter.

If hay is being fed in a field shelter away from the house it is far better to put it loose on the ground rather than run the risk of an animal becoming entangled in a hay-net. Feeding hay in this fashion is not the most economical method, but it is certainly preferable to having a foal so badly injured that it may have to be destroyed.

The greatest temptation connected with breeding a foal is that of starting it in work, but *on no account* should a youngster be asked to carry weight until it is three years old, or more preferably three and a half. Lessons in hand should of course continue, but absolutely no riding.

After the first winter there is no reason why yearlings and two-year-olds should not live out night and day throughout the year, although they will, of course, still require good supplementary feeding. Good grazing may suffice throughout the summer months, but they will need as much good hay as they will consume during the winter as well as a daily concentrate ration. It can be taken as a rough guide that an animal likely to mature to 14.2 h.h. will need a concentrate ration of 6 lb. or 7 lb., the smaller ponies taking 1 lb. to 2 lb. less and the larger, hunter type, youngsters needing 2 lb. to 3 lb. more. Worm control must be continued; feet will need attention every four to six weeks; and it is advisable to ask a veterinary surgeon to inspect the youngster's teeth once or twice a year, as young horses need to have their molars rasped just as the older animals do. A tetanus toxoid booster dose will be required annually, and most owners will want to protect their horses and ponies from equine influenza, for which purpose there is a vaccine available.

Careful attention to such details, in addition to proper handling and good feeding, will build a future mount of whom you will have reason to be proud, and when, in a further two and a half to three years you are able to continue educating your home-bred youngster for the saddle, the pleasure and satisfaction you will gain will be a worthwhile reward for the years of waiting.

Appendix

BRITISH SOCIETIES

Arab Horse Society: Lough Moe, Itchen Abbas, Winchester, Hampshire

British Palomino Society: Kingsettle Stud, Cholderton, Salisbury, Wiltshire

British Show Pony Society: Smale Farm, Wisborough Green, Sussex

British Spotted Horse Society: Nash End, Bisley, Stroud, Gloucestershire

Dales Pony Society: Hutton Gate, Guisborough, N. Yorkshire

Dartmoor Pony Society: Lower Hisley, Lustleigh, Newton Abbot, Devon

English Connemara Pony Society: The Quinta, Bentley, Farnham, Surrey

Exmoor Pony Society: Capland Orchard, Hatch Beauchamp, Taunton, Somerset

Fell Pony Society: Packway, Windermere, Westmorland

Highland Pony Society: Dunblane, Perthshire

Hunters' Improvement Society: 17 Devonshire Street, London W.1

National Pony Society: Stoke Lodge, 85 Cliddesdon Road, Basingstoke, Hampshire

New Forest Pony and Cattle Breeding Society: Beacon Corner, Burley, Ringwood, Hampshire

Ponies of Britain Club: Brookside Farm, Ascot, Berkshire

Shetland Pony Stud Book Society: 8 Whinfield Road, Montrose, Angus

Welsh Pony and Cob Society: 32 North Parade, Aberystwyth, Cardiganshire

INDEX